Wild & Furry Animals
of the
SOUTHERN APPALACHIAN MOUNTAINS

Written and Illustrated
by
Lee James Pantas

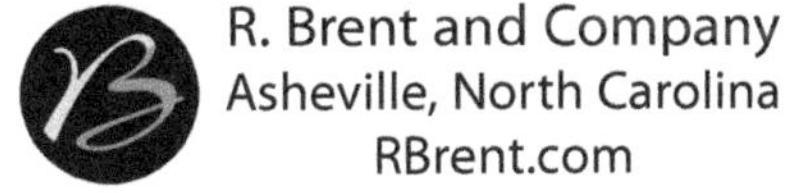
R. Brent and Company
Asheville, North Carolina
RBrent.com

Published in Asheville, North Carolina
by R. Bren[illegible]d Company
50 Deerwood Dr.
Asheville NC [illegible]05
828-231-3923
robbin@rbrent.c[illegible]

Editor and publisher: *Robbin Brent*
Cover design: *Jane War[illegible]graphics*
Design & layout: *Rick So[illegible]*
Illustrations: *Lee James Pant[illegible]*

Publication Data

Pantas, Lee James
wild & furry animals of the southern appalachian mountains
p. cm.
ISBN Casebound: 978-0-9910398-[illegible]-8
ISBN Paperback: 978-0-9910398-2-1

Lee James Pantas may be contacted at 828-779-1569 or leepantas@bellsouth.net for ordering information. Visa and MasterCard accepted. This book is also available for order online at leepantas.com

*This book is dedicated to my beloved wife, Elizabeth,
the original Mrs. Otter!*

Contents

Introduction

The ancient mountains of the Southern Appalachians are a vast and wonderful place, and are home to over a hundred species of extraordinary and interesting wild mammals. When I first moved to the beautiful Fairview valley near Asheville North Carolina where I live, I realized right away that in the deep woods and lush green fields surrounding my house there were numerous gray squirrels, rabbits and other small animals, and occasionally black bears, bobcats and white-tailed deer. As time passed I discovered other even more secretive and mysterious creatures, including an industrious muskrat couple living down by the stream, trout-eating mink in residence under the bridge, a mystical black coyote that occasionally passed through my wife's rose garden on its way to my neighbor's chicken coop, and a flying squirrel colony living in an old tree. Over the years, this list of wild animals I have encountered continued to grow and includes little brown bats that now occupy the pond-facing eaves of our house, an acrobatic woodland jumping mouse that left me amazed while I saw it bounding along like a miniature kangaroo, and a resourceful little field mouse that I met eye to eye when I was cleaning out a bluebird box, evidently having decided this was its perfect new home. Inspired by these recent encounters of the furry kind, last year I decided to document by illustration and text some of the more common as well as some of the lesser known creatures of this wonderful Appalachian realm, and with that decision this book was born.

An act of kindness towards an animal is a step closer to God.

About the Author

Lee James Pantas, born in 1941, has lived in the Asheville NC area since 1989. An artist and author, he also has a master's degree in Ecology (University of Vermont, 1966) and is active in environmental causes.

In 1978 he founded a Catholic soup kitchen New Covenant House in Stamford Ct. to feed homeless people. Since then it has grown into an 8000 sq. ft. center now known as New Covenant Center that serves over 700,000 meals a year to needy persons along with a wide range of other social services.

Mr. Pantas has also coached track & field for the A.C. Reynolds Middle and High Schools for over 25 years and has established a reputation as one of the top high school hurdles coaches in America. In 2003 he co-founded the Blue Ridge Classic and has helped grow this annual event into the largest high school track and field meet in North Carolina.

He has completed more than 2,000 architectural pen & ink drawings of private homes as well as more than 500 historical buildings, inns, churches, and animals. He is also the author of The Ultimate Guide to Asheville & the Western North Carolina mountains, a best-selling illustrated travel guide.

As a painter, he is well known for his visionary and fantasy paintings. His work has been exhibited in galleries nationwide and is found in many private and public collections. In 2014 he completed a series of imaginary coral reef paintings entitled Liza's Reef to draw attention to the plight of oceans worldwide caused by pollution, overfishing and global warming.

Correspondence should be directed to:

Lee James Pantas
18 Garren Mountain Lane
Fairview NC 28730

leepantas@bellsouth.net
828-779-1569

Websites: cherryorchardstudio.com
leepantas.com
ashevilleguidebook.com
lizasreef.com

Wild & Furry Animals of the Southern Appalachian Mountains

The wild and furry animals presented in this book are found in a vast and beautiful region of America referred to as the Southern Appalachian Mountains. The eight-state region includes the Allegheny Mountains of West Virginia and Virginia; the Blue Ridge Mountains of Virginia, western North Carolina, northwestern South Carolina and the northeastern Georgia; the Unaka Mountains of southwestern Virginia, eastern Tennessee and western North Carolina; and the Cumberland Mountains of southwestern West Virginia, southwestern Virginia, eastern Kentucky, eastern Tennessee and northern Alabama.

This book, written and illustrated by Asheville pen & ink artist Lee James Pantas, presents all of the mammals, both large and small that are likely to be encountered in the mountains, as well as the some of the rarest and least likely to be seen.

American Black Bear

Ursus americanus

What is their preferred habitat?

Uninhabited forests.

What do they like to eat?

They are omnivorous and eat a wide variety of foods, including acorns, roots, insects, plants, berries, fruit, fish and small mammals. They also will kill and eat young deer, elk or sheep.

What is their average life span in the wild?

10 to 20 years.

How big are they?

Large males weigh between 100 and 500 pounds but can grow to 600 pounds. Smaller females typically do not exceed 200 pounds.

What is their fur color?

Usually black but also brown and dark brown.

Fun Facts

- American black bears are found only in North America and are the smallest of the three bear species found on the continent.
- They have short, non-retractable claws that give them excellent tree-climbing abilities. They are highly dexterous and capable of opening door latches and screw-top jars.
- Most hibernate in the winter depending on local weather conditions and food supply.
- Black bears can easily outrun humans, with tops speeds of 25–30 miles per hour. They can run uphill and downhill with equal speed.
- Black bears are good swimmers and will readily enter water in search of fish.
- Their sense of smell is much greater than that of dogs.
- Some black bears in British Columbia are pure white in color, and are referred to as "Spirit Bears" by the local Native Americans.

American Mink

Neovison vison

What is their preferred habitat?

Forested areas near rivers, streams, marshes, swamps, ponds and lakes.

How big are they?

Males are typically 13 to 18 inches long and weigh from 1 to 3 pounds with female minks slightly smaller.

What do they like to eat?

Favorite prey are fish but they also feed on crustaceans, amphibians, rodents and birds.

What is their average life span in the wild?

1 to 3 years.

What is their fur color?

Dark brown to black.

Fun Facts

- American minks are graceful swimmers that can dive as deep as 16 feet under water. They swim by using their back legs.
- They run on land in a distinctive bounding gait and can climb trees with ease.
- Minks are solitary creatures.
- Minks are usually able to catch fish after a furious 5-to 20-second chase.
- James Audubon, the famous naturalist, once reported seeing a mink carrying a 12-inch trout.
- They kill snakes but do not eat them.

Bobcat

Lynx rufus

What is their preferred habitat?

Woodlands

How big are they?

Bobcats are about twice as big as an ordinary housecat and generally weight between 11 and 28 pounds, with females smaller than males. There have been rare reports of bobcats reaching 60 pounds.

What do they like to eat?

Rabbits and hares are #1 on their list but they also eat birds, mice, and many other small animals.

What is their average life span in the wild?

7 years.

What is their fur color?

Beige to brown with spotted or lined, dark brown or black, markings.

Fun Facts

- Bobcats are named for their short, bobbed tails.
- Bobcats are nocturnal, solitary animals and are rarely seen by humans. Their growls are deep and fearsome, sounding as if they were coming from a much larger animal.
- They are fierce stealth hunters and can kill animals much bigger than themselves, including deer.
- They can leap over 10 feet and are excellent swimmers.
- They are the most abundant wild cat in America and can have home ranges as large as 20 square miles.

Cougar (Mountain Lion)

Puma concolor

What is their preferred habitat?

Forests with dense underbrush and rocky areas.

How big are they?

Between 5 to 9 feet long from head to tail. Males can weigh up to 150 pounds while smaller females can reach 100 pounds.

What is their average life span in the wild?

10 to 14 years.

What do they like to eat?

Favorite prey are deer but they also feed on smaller animals as well as wild boar, domestic cattle, and sheep.

What is their fur color?

Tawny brown to reddish with lighter colors on underbelly. Their tails have a black spot on the tip.

Fun Facts

- Officially considered extinct in the Southern Appalachian Mountains by 1900. Even though numerous sightings have occurred over the years, these are not considered proof the mountains have a resident population. Sightings may be from released captive cougars, large domestic cats or wide-ranging individuals from other areas.
- Young male cougars in search of new territory away from other more dominant males have been known to travel thousands of miles. In 2011, a cougar from the Black Hills of South Dakota was hit and killed by a car in Greenwich, Connecticut, on the Wilbur Cross Parkway. Determination of the cat's origin from a population in South Dakota was confirmed through DNA analysis.
- Cougars cannot roar but they do scream and can purr like a house cat.
- Cougars have no natural enemies and are at the top of the food chain. They feed just once or twice a week.
- They are ambush predators with top bursts of speed 35 to 45 miles per hour.
- Cougars have large paws and proportionally the largest hind legs in the cat family. These allow for great leaping ability. An amazing vertical leap of 18 feet and horizontal jumps of up to 40 feet have been reported.
- Cougars often stalk prey for up to an hour.
- Other names for cougars are Mountain Lion, Puma, Panther and Catamount. The cougar holds the title in the Guinness Book of World Records for the animal with the greatest number of names, with over 40 in English alone.
- The city of Cusco in Peru is reported to have been designed in the shape of a cougar.

Coyote
Canis latrans

What is their preferred habitat?
Woodlands

How big are they?
Between 20 and 50 pounds.

What is their fur color?
Light gray to reddish brown with areas of black and white.

What do they like to eat?
They are omnivores and will eat almost anything. They actively hunt rabbits, fish, frogs, rodents, and other small mammals as well as insects, snakes and fruit.

What is their average life span in the wild?
4 to 6 years.

Fun Facts

- Coyotes are one of the most adaptable and cunning canines on the planet and can be found in a wide variety of habitats, from deserts to alpine meadows, and even in large cities like New York and Los Angeles.
- They are social animals and prefer to live and hunt in packs.
- Coyote howls, referred to by scientists as group yip-howls, are short howls that rise and fall in pitch, punctuated with staccato yips, yaps, and barks. These group yip-howls are often mistaken for a large pack of animals all raising their voices at once. Because of the variety of sounds produced by each coyote, and the way sound is distorted as it passes through the environment, two coyotes can sound like six to eight animals.
- When walking or running, the coyote carries its tail downwards rather than horizontally as wolves do.
- They are very strong swimmers.
- Coyotes can and do breed with wolves and dogs.
- Coyotes raise their pups in dens or holes in the ground.

LEE JAMES PANTAS

Deer Mouse

Peromyscus maniculatus

What is their preferred habitat?

Woodlands and meadows with plenty of ground cover.

How big are they?

Small, their bodies are only 3 to 4" long.

What is their average life span in the wild?

1 year.

What do they like to eat?

They are omnivorous and eat primarily seeds, flowers, nuts, fruits, leaves, fungi and insects, including spiders and caterpillars.

What is their fur color?

Grayish brown with white underbellies.

Fun Facts

- Deer mice are accomplished climbers, jumpers, and runners, and their name "deer mice" refers to their athletic abilities as well as their fur coloration, which is like that of deer.
- They are nocturnal creatures and spend their days hidden in the nests they build in burrows, hollow trees, or logs.
- Female deer mice construct the nests from grasses, roots, wool, thistle down, moss, and a variety of other natural materials.

Eastern Chipmunk

Tamias striatus

What is their preferred habitat?

Forests and woodlands with plenty of ground cover.

How big are they?

Small, their bodies are only 2 to 6" long.

What is their average life span in the wild?

2 to 3 years.

What do they like to eat?

They are omnivores with their diet consisting primarily of seeds, fungi, nuts, fruit and grass.

What is their fur color?

Reddish-brown on its sides and white on the stomach, with black and white striping on sides and back.

Fun Facts

- Chipmunks have cheek pouches that allow them to carry large amounts of food items to their burrows for storage and consumption. One chipmunk was observed gathering more than 160 acorns in one day. In just two days a chipmunk can collect enough food to last for the winter.
- Chipmunks have been known to store up to 8 pounds of food in their burrows.
- Chipmunks dig extensive burrows underneath or next to natural or manmade cover, including logs, stumps, and rocks.
- Chipmunks are curious animals and can be coaxed into eating out of a human hand.
- A group of chipmunks is called a scurry.
- North America is home to 21 chipmunk species.
- Chipmunks hibernate in the winter but don't sleep all the way through the season. They retreat to their burrows in late fall but wake every few days, raise their body temperatures to normal, feed on stored food, and then go back to sleep.

Eastern Cottontail Rabbit

Sylvilagus floridanus

What is their preferred habitat?

Meadows, open grassy areas, forest clearings and old fields with abundant grass and plants for food and shrubs for cover.

What do they like to eat?

They are herbivores and eat a variety of grasses, clover, leaves, flowers, bark and seeds.

How big are they?

Between 1 and 3 pounds.

What is their average life span in the wild?

1 to 2 years.

What is their fur color?

Speckled brown-gray and reddish brown.

Fun Facts

- Eastern cottontail rabbits do not dig burrows, but instead nest in shallow, slanted holes that they dig under clumps of grass or shrubs and then line with grass and fur.
- They are very active animals, but also can remain sitting completely still for 15 minutes at a time.
- Cottontail rabbits thrive as a species because they reproduce up to five times per season. A single pair of rabbits could potentially produce 350,000 rabbits in just five years if there were no predators.
- They can leap up to 15 feet when necessary to escape predators. When pursued, they circle their territory and jump sideways to break their scent trails.
- Their name is derived from their white, fluffy tail that resembles a ball of cotton.
- On midwinter nights, cottontail rabbits have been observed frolicking on crusted snow. These playful gatherings are believed to be reactions after periods of forced inactivity.

Eastern Gray Squirrel

Sciurus carolinensis

What is their preferred habitat?

Forests and woodlands, preferably those with oak, walnut, hickory, and other nut-bearing trees.

What do they like to eat?

Acorns, nuts, seeds, tree buds, bark, and any food found in residential bird feeders.

How big are they?

1 to 2 pounds.

What is their average life span in the wild?

8 to 9 years.

What is their fur color?

Gray

Fun Facts

- Eastern gray squirrels are easily distinguished from the larger fox squirrels by their predominantly gray color. Gray squirrels also have both black and white-colored variations. Genetic variations also include individuals with black squirrels with white tails, and gray colored squirrels with black tails.
- Gray squirrels are scatter-hoarders, animals that store food in numerous caches for later recovery during winter months. Individual squirrels can make up to several thousand caches each season. They often lose up to 25% of their buried food to fellow squirrels or birds, and they also don't find all of their buried nuts, which results in the growth of more trees.
- Squirrels sometimes use deceptive behavior to prevent other animals from finding their food caches by only pretending to bury a food item when they think they are being watched.
- Squirrels can find food buried beneath a foot of snow. Once they find a scent to buried food, they dig a tunnel through the snow to the buried treasure.
- Most squirrel populations in major city parks were introduced by humans.
- There are over 265 species of squirrels worldwide. The smallest is the African pygmy squirrel only a few inches long, to the Indian giant squirrel, which measures three feet long.
- They are one of the few mammals that can descend a tree head first.
- When threatened, they run away in a zigzag pattern. This is an extremely useful strategy in escaping predators but unfortunately does not work well when they are on roads trying to avoid being run over by cars.
- They construct their dens as nests from twigs and leaves in tree branches, or within the hollow trunks of trees.

LEE JAMES PANTAS

Eastern Mole

Scalopus aquaticus

What is their preferred habitat?

Eastern moles prefer moist, loose soils found in fields, meadows, pastures, and wooded areas.

What do they like to eat?

Earthworms but they also will eat many other types of food, including snails, centipedes, insects, grubs, and slugs.

How big are they?

Typical size is around 6 inches in length and they usually weigh only a couple of ounces.

What is their average life span in the wild?

6 years.

What is their fur color

Dark to light shades of silver gray.

Fun Facts

- Eastern moles spend 99% of their time underground.
- They have tiny, degenerative eyes with fused eyelids that are limited to distinguishing only between shades of light and dark.
- They have enormous hairless front feet with curved claws that are used for digging.
- They dig burrows just under the surface for foraging. When digging these burrows, they push excess soil up onto the surface referred to by scientists as "molehills."
- Moles can dig tunnels at a rate of more than 15 feet per hour.
- One of the mole's favorite habitats are golf courses.
- Each day moles must eat 25 % to 100% of their own weight in food to survive. One mole can eat over 50 pounds of worms in a year.
- Moles can smell in stereo. This ability helps them to not only to detect odors, but also to determine the direction from which the odors are coming.

Eastern Spotted Skunk

Spilogale putorius

What is their preferred habitat?

Forests and wooded areas.

How big are they?

They are smaller than Striped skunks, and weigh usually between one and two pounds.

What is their average life span in the wild?

1 to 2 years.

What do they like to eat?

They are omnivores and prefer rabbits, small rodents, insects, birds, fruit, and eggs.

What is their fur color?

Black and white.

Fun Facts

- Eastern spotted skunks are rare in the Southern Appalachian Mountains and not much is known about their distribution in the mountains.
- Eastern spotted skunks have 4 to 6 stripes on their backs, giving them a spotted appearance. The only true spot they have is one found on their foreheads. Their distinctive black and white appearance is a warning to potential predators.
- They are excellent climbers and can scurry up and down trees like squirrels.
- Before spraying a potential predator, they stand repeatedly on their front paws in a handstand as a warning display. If that doesn't work, they then assume a horseshoe shaped stance before spraying their foul-smelling musk towards the predator. They have an accurate range of over 10 feet.
- They are nocturnal animals and are usually seen by humans only at dawn or dusk.

Elk

Cervus canadensis

What is their preferred habitat?
Forests and meadows.

How big are they?
Elk cows average between 500 to 530 pounds, while larger bulls average from 700 to 730 pounds. Mature bulls stand 5 feet high at the shoulder, and are up to 8 feet in length.

What do they like to eat?
Grasses, plants, leaves, and bark.

What is their average life span in the wild?
10 to 13 years.

What is their fur color?
In summer, copper brown, and light tan throughout the rest of the year.

Fun Facts

- Elk were eliminated by 1810 in the Southern Appalachian Mountains of North Carolina, Tennessee, and Kentucky by over hunting and loss of habitat. In 1997, however, they were successfully reintroduced into Kentucky, and later in 2001 into North Carolina and Tennessee in the Cataloochee Valley. These areas now have thriving herds.
- The Cataloochee Valley, located in the Great Smoky Mountains National Park, is one of the more popular viewing sites for those who want a close-up glimpse of these magnificent creatures in the wild.
- Elk, also known as Wapiti, are one of the largest species in the world-wide deer family.
- They live in herds that are defended and dominated by one bull.
- They have an amazing vertical leap of 8 feet and easily jump over fences.
- Only male elk have antlers, which are shed each winter. Antlers may reach 4 feet in length and in summer can grow more than an inch a day.
- Elk can outrun horses, with top speeds of 45 miles per hour.
- Bulls have a distinctive vocalization known as "bugling," which can be heard for miles. These calls are used mainly during mating season and females are attracted to the males that bugle most often and loudest.
- The major predators of elk are not wolves but bear and mountain lions.
- The journals of explorers Lewis and Clark were bound in elk hide.
- The species of elk currently living in the Southern Appalachian Mountains, are small compared to those that once roamed the region. These forerunners of modern elk had antlers 6 feet long and weighed over 1,000 pounds.

Fox Squirrel

Sciurus niger

What is their preferred habitat?

Forests and woodlands, especially those with oak, hickory, walnut, and pine trees.

How big are they?

Usually between 17 and 27 inches long, with an 8 to 13 inch tail. They typically weigh around 2 pounds.

What is their average life span in the wild?

3 to 4 years.

What do they like to eat?

Favorite foods are acorns, other types of nuts, seeds, tree buds, bulbs, roots, and insects.

What is their fur color?

Gray to tawny brown, with reddish-brown to pale gray underneath.

Fun Facts

- Fox squirrels are the largest species of tree squirrels native to North America.
- Fox squirrels have strikingly-patterned dark brown and black fur with white bands on the face and tail.
- They have two types of shelters that they use: tree dens and leaf nests. They prefer to use the tree dens during the winter and for raising young because of the better insulation the dens provide.
- Great climbers and jumpers, they can leap over 15 feet in horizontal jumps from one branch to another.
- There are more than 265 species of squirrels worldwide. The smallest is the African pygmy squirrel, only a few inches long, to the Indian giant squirrel, which measures three feet long.
- Squirrels can find food buried beneath a foot of snow. Once they find a scent to buried food, they dig a tunnel through the snow to the buried treasure.
- Like gray squirrels, fox squirrels bury nuts in the ground to feed on during the winter.
- Fox squirrels are scatter-hoarders, animals that store food in numerous caches for later recovery, especially during winter months. Individual squirrels can make up to several thousand caches each season.
- They may lose up to 25% of their buried food, which may result in the growth of more trees.
- They are one of the few mammals that can descend a tree head first.
- When threatened, they run away in a zigzag pattern. This is an extremely useful strategy in escaping predators but unfortunately does not work well when they are on roads trying to avoid being run over by cars.
- They construct their dens as nests from twigs and leaves in tree branches, or within the hollow trunks of trees.

Gray Fox

Urocyon cinereoargenteus

What is their preferred habitat?

Habitats with lots of woods and brush.

How big are they?

Larger than the red fox, gray foxes typically weigh between 7 and 15 pounds.

What is their average life span in the wild?

6 to 8 years.

What do they like to eat?

They are omnivorous and their favorite prey is the Eastern cottontail rabbit. They also eat voles, shrews, birds, crickets, grasshoppers, and fruit.

What is their fur color?

Peppery-gray on their backs, and reddish-brown on chest, sides and legs.

Fun Facts

- The gray fox's range includes both North and South America.
- They have the ability to climb trees, shared only with the Asian raccoon dog among canids. They have strong claws that allow them to climb to escape predators, including coyotes and domestic dogs, or to access arboreal food sources. They also can jump easily from branch to branch.
- Gray foxes sometimes take naps in trees.
- They make dens in hollow trees, stumps, or burrows to hide from predators and to raise their young.
- Gray foxes are easily differentiated from the red fox by their lack of "black stockings" on their lower legs, and have predominately gray fur, not orange.
- They have vertical pupils and can retract their claws, two physical characteristics of cats.
- Female foxes are called vixens.
- Foxes have whiskers on their face and legs, which helps them navigate.
- They have superior hearing and have been reported to be able to hear a watch ticking over 40 yards away.
- Gray foxes are nocturnal and hunt primarily at night.

Groundhog (Woodchuck)

Marmota monax

What is their preferred habitat?

Open meadows and the edges of woodlands.

How big are they?

Usually between 4 to 13 pounds and have been recorded to weigh over 30 pounds.

What is their average life span in the wild?

3 to 6 years.

What do they like to eat?

Wild grasses and other vegetation. Alfalfa, dandelion, clover, and coltsfoot are some of their favorite grasses.

What is their fur color?

Grayish brown.

Fun Facts

- Groundhogs are excellent swimmers and are related to squirrels.
- They can climb trees.
- They have two coats of hair that gives them a distinctive "frosted" appearance.
- The largest Groundhog Day celebration is held in Punxsultawney, Pennsylvania, the home of its famous resident groundhog, Punxsultawney Phil.
- Groundhogs dig extensive burrows, sometimes over 60 feet long, which are used for hibernating in the winter, sleeping, and nurseries, as well as a safe retreat when danger is present. There are usually 2 or more entrances.
- February 2nd in the United States and Canada is celebrated as Groundhog Day. According to tradition, if it is sunny when it emerges from its burrow, the groundhog will supposedly see its shadow and retreat back into its den, and the winter weather will persist for 6 more weeks. If it is cloudy and it sees no shadow, then the spring season will arrive early.
- One of their most common nicknames, "woodchuck" has nothing to do with wood. It's thought to be a corruption of the Native American words wejack, woodshaw, or woodchoock.

Least Weasel

Mustela nivalis

What is their preferred habitat?

Least weasels are adaptable and thrive in a multitude of habitats from forests to meadows.

What do they like to eat?

Primarily small rodents as well as occasionally rabbits, birds, fish, and frogs.

What is their average life span in the wild?

1 to 2 years.

How big are they?

Least weasels are the smallest of the weasel family and usually are only 8 to 9 inches long and weigh only a few ounces.

What is their fur color?

For most of the year, light brown with a white underbelly. In winter, their fur changes to all white.

Fun Facts

- Least weasels are rare in the Southern Appalachian Mountains and not much is known about their distribution in the mountains.
- The least weasel is the smallest carnivorous predator in the world. Despite its small size, the Least weasel is a fierce hunter and capable of killing a rabbit ten times its size.
- Because of its size and very active lifestyle, the Least weasel must eat 40 to 60% of its body weight every day.
- This white winter fur glows a bright lavender color under ultraviolet light.
- Least weasels are solitary animals and make nests in crevices and roots of trees, and abandoned burrows.
- In killing its prey, the weasel's method is like that of a jaguar. After jumping on its prey, it wraps its muscular body around the animal to immobilize it and then delivers a single fatal bite to the back of the head, puncturing the skull or spinal cord.
- They will hunt and kill anything that moves and look like prey, even if they are fully fed. They often kill much more than they can eat. Any carcasses not consumed are stored in in underground caches near their dens to eat at a later time.
- The coats of the Least weasels living in the northern part of its range turns white in the winter to blend in with the snow. This is characteristic of all weasels.

LEE JAMES PANTAS

Little Brown Bat

Myotis lucifugus

What is their preferred habitat?

Near swamps, ponds, lakes, rivers, and streams.

How big are they?

Very small, usually between 2.5 and 4 inches long. They also weigh no more than an ounce.

What is their average life span in the wild?

6 to 7 years.

What do they like to eat?

Insects, especially aquatic insects, gnats, midges, flies, moths, wasps, and beetles.

What is their fur color?

Uniformly brown.

Fun Facts

- The little brown bat is one of the most common bats in North America.
- Evening forages as a group are done usually above bodies of water.
- Little brown bats sleep and groom during the day and hunt at night, usually only for a few hours.
- Bats use echolocation to find their prey.
- They have three different roosting sites: day roosts and night roosts during spring, summer, and fall; and hibernation roosts in the winter. Day roosts are typically found in trees or buildings, wood piles, and occasionally caves. Night roosts tend to be in the same area as day roosts but are in tighter, closed-in spaces so the bats can huddle together for warmth.

Meadow Vole (Field or Meadow Mouse)

Microtus pennsylvanicus

What is their preferred habitat?

Moist grassland areas with thick protective cover.

How big are they?

Usually between 5 to 8 inches long, including the tail.

What is their average life span in the wild?

Less than 1 year.

What do they like to eat?

Grasses, sedges, and other vegetation, including many agricultural plant species.

What is their fur color?

Dark brown above and silvery-gray below.

Fun Facts

- Meadow voles are one of the most common and prolific small mammals in North America.
- They can swim but have never been seen climbing.
- Depending on the type of habitat, the local population of meadow voles varies widely, from only a few individuals an acre to more than 1,480 per acre.
- Hawks sitting on telephone and electrical wires overlooking pastures are often watching for voles below.
- Meadow voles are prolific breeders and can have up to 17 litters a year, with the average litter size between 4 to 8 babies.

Muskrat

Ondatra zibethicus

What is their preferred habitat?

Wetlands and swamps, and along streams, rivers, canals, lakes, and ponds.

How big are they?

Usually between 2 and 5 pounds and 1 to 2 feet long.

What is their average life span in the wild?

1 to 3 years.

What do they like to eat?

Cattails, water lilies, and other vegetation. In winter they will also each insects, crayfish, fish, and freshwater mussels.

What is their fur color?

Medium to dark brown or almost black.

Fun Facts

- Muskrats are not related to rats although their appearance is similar. They are named for their musky scent, which is emitted from glands at the base of their tails.
- Muskrats are semi-aquatic and spend much of their time in water. They are great swimmers and can swim underwater, backwards and forwards, for over 15 minutes. Their main means of propulsion in the water are their flattened, black scaly tails, which function as rudders.
- For shelter, muskrats build burrows into banks with underwater entrances along streams and lakes. If they live in marshes and swamps, they build 3-to 5-feet-high lodges made of vegetation and mud, also with underwater entrances.
- The inner chamber of burrows and lodges always rise above the surface of the water to give the muskrats dry shelter.
- Muskrats do not hibernate and are active year-round, feeding at all times of the day.
- Muskrats gather food and carry it to feeding platforms that they build for a place to eat. These platforms are flat, elevated piles of mud and vegetation constructed outside of their burrows. If abandoned, these platforms also are used as nesting sites by many birds
- Most of the time muskrats mate underwater, and their social life is built around the colony, which typically consists of an adult pair and 2 to 6 young.

North American Beaver

Castor canadensis

What is their preferred habitat?
Swamps, marshes, streams, lakes and ponds.

What do they like to eat?
Leaves, buds and inner bark of live trees, especially aspens, poplar, beech, birch, willow, and alder.

What is their average life span in the wild?
10 to 20 years.

How big are they?
They typically weigh between 20 and 70 pounds. Very old individuals have been recorded at over 100 pounds.

What is their fur color?
Chestnut brown to almost black.

Fun Facts

- The beaver is the largest rodent in North America and is the 2nd largest in the world.
- Beavers do not eat fish.
- The large front teeth of the beavers, used constantly for gnawing trees and branches, never stop growing. These teeth are orange due to the iron contained in the enamel, a substance that also makes them incredibly strong and sharp.
- Beavers are best known for constructing dams across steams to create ponds. The dams are constructed from mud, rocks, vegetation, and branches from trees the beavers cut down. They also construct lodges equipped with underground entrances in the middle of the ponds created by their dams.
- Beavers are semi-aquatic and are excellent swimmers. They can stay submerged for up to 15 minutes and much prefer to be in the water rather than on land.
- They are primarily nocturnal animals and do much of their dam building at night.
- Beavers like to stay busy and are known as prolific builders. Hence the saying, "As busy as a beaver."
- During the Ice Ages, beavers, known as Castoroides, grew up to 8 feet long and weighed over 200 pounds.
- Their dams can be enormous structures and one dam, under construction since the 1970s in northern Alberta, measured over 2,500 feet long.
- Beavers were brought to the brink of extinction by 1838 in North America because of their fur, which was highly sought after for clothing and hats. Much of the westward expansion in America was driven by the quest for beaver fur. With protection in the late 19th and early 20th centuries, the beaver population in America has rebounded to around 15 million.

North American River Otter

Lontra canadensis

What is their preferred habitat?
Lakes, rivers, streams, swamps, and marshes.

How big are they?
Between 11 and 30 pounds.

What is their average life span in the wild?
8 to 9 years.

What do they like to eat?
They are carnivores and their primary foods are fish, crayfish, amphibians, clams, mussels, snails, and turtles.

What is their fur color?
Light brown to black.

Fun Facts

- River otters swim by propelling themselves with their powerful tails and flexing their long bodies and webbed feet. They can hold their breath underwater for up to 8 minutes.
- Otters are renowned for their sense of play and love to slide down snow-covered or muddy banks into the water. They are equally at home on land or in the water.
- Unlike most marine mammals, river otters do not have a layer of insulating blubber. Instead, air is trapped in their dense fur, which helps to keep them warm.
- They are extremely social animals and form large social groupings, typically from 4 to 8 individuals, with some groups as large as 20.
- They live in dens, constructed in the burrows of other animals, or in naturally occurring hollows under logs or river banks.
- They have specialized vision that enables them to view potential meals while swimming in dark or murky water. They easily spot food not visible to other animals. They possess an exceptional sense of smell that also helps them locate prey.

Northern Flying Squirrel

Glaucomys sabrinus coloratus & Glaucomys sabrinus fuscus

What is their preferred habitat?

Coniferous and mixed coniferous forests.

How big are they?

About 10 inches from their nose to the tip of their tail.

What is their average life span in the wild?

4 to 6 years.

What do they like to eat?

Primarily acorns, nuts, fungi and lichens. They also will eat tree sap, insects, bird eggs, and nestlings.

What is their fur color?

Light brown or cinnamon with creamy white fur underneath.

Fun Facts

- In the Southern Appalachian Mountains, Northern flying squirrels are rare and found only in a few higher elevation areas in Virginia, North Carolina, and Tennessee.
- There are two subspecies in the Southern Appalachian Mountains: the Carolina Northern flying squirrel (G. s. coloratus) in southwestern Virginia and western North Carolina, and the Virginia Northern flying squirrel (G. s. fuscus) in northern Virginia, both of which are endangered.
- Northern flying squirrels do not actually fly, but instead glide by opening a membrane called a patagium to catch the air. They have been observed to glide up to 75 feet from tree to tree, with their average glide around 15 feet.
- They are strictly nocturnal and generally nest in the holes of trees, preferring dead trees of larger diameter.
- It is believed that they use triangulation to estimate the distance of the landing area when they launch themselves into the air.
- They are social animals and have been observed flying and foraging together in large groups.
- Size is one way to tell the Southern and Northern flying squirrels apart. The Southern flying squirrels average 8 to 10 inches, while the northern variety are usually 10 to 12 inches long.
- Humans imitate flying squirrels using special suits called "wingsuits," which are designed after the squirrel's anatomy with its extended air-catching flaps of skin. These wingsuits allow the experienced wearer, jumping from high mountainous elevations and off cliffs, to glide just like the squirrels over long distances.

Northern Short-Tailed Shrew

Blarina brevicauda

What is their preferred habitat?

Grasslands, marshy area, gardens and forests, especially those with moist with leaf litter or thick plant cover.

What do they like to eat?

Primarily carnivorous, they eat insects, earthworms, voles, mice, snails, salamanders, and other shrews.

How big are they?

Between 4 to 5 inches long, and weighing around an ounce.

What is their average life span in the wild?

1 to 2 years.

What is their fur color?

Silvery slate gray, occasionally black or brownish black.

Fun Facts

- The Northern short-tailed shrew is notable in that it is one of the few venomous mammals in the world. Their saliva contains a chemical that paralyses their prey and is toxic enough to kill small animals several sizes larger than the shrew itself. Their saliva is not strong enough to kill a human, but will cause a very painful bite.
- They are voracious eaters and must consume up to three times their body weight each day.
- In the winter, shrews are active in the airspace between the ground and snow cover.
- Shrews hunt using echolocation and a developed sense of touch as they tunnel along below ground, and through snow or leaf litter.
- Their eyes are degenerated and vision is thought to be limited only to the detection of light.
- Nests lined with fur and leaves are built in underground chambers or under logs or rocks.
- A small pile of old snail shells on the ground might be a tell-tail sign of a shrew nest.

Norway Rat (Brown Rat)

Rattus norvegicus

What is their preferred habitat?

Almost exclusively in areas near human habitation.

How big are they?

Usually around 10 inches long and between 10 and 16 ounces.

What is their average life span in the wild?

1 year or less.

What do they like to eat?

They are omnivores and will eat almost anything. Cereal grains and seeds are a substantial part of their diet.

What is their fur color?

Brown to dark gray.

Fun Facts

- Norway rats are one of the best known and most common rats in the world. By population and territory occupied, they qualify as the second most successful mammal on the planet after humans.
- They live wherever humans live, especially in urban areas. It is estimated there is one rat for every person in the United States.
- Researcher Martin Schein found that the most-liked foods of Norway rats include macaroni and cheese, raw carrots, scrambled eggs, and cooked corn kernels.
- Romans considered the rat to be a sign of good luck.
- The rat is the first symbol of the Chinese Zodiac and stands for cunning and prosperity.
- The ancient Mayan and Egyptian civilizations worshiped the rat.
- An adult rat can jump three feet vertically into the air, and over four feet horizontally from a standing position.
- Rats sleep together and commonly groom each other. They dig extensive burrows with multiple levels of tunnels when soil conditions permit that provide shelter, nest areas, and food storage sites. They have a very powerful chain of command and the largest and strongest rats get the best food and shelter places.
- Karni Mata Temple is a Hindu temple dedicated to Karni Mata in Rajasthan, India, known as the Temple of Rats. It is famous for the approximately 20,000 black rats that live and are fed by the temple staff in the temple and surrounding area.
- Rats were determined to be part of the means by which The Plague was spread in Europe in the 1300s when millions of people lost their lives. Rats were the host carriers to fleas, which in turn carried the bacterium, Pastuerella pestis, the cause of the disease. Both humans and rats died from infection by the bacteria.

Raccoon

Procyon lotor

What is their preferred habitat?

Forests and woodlands.

How big are they?

Between 8 and 23 pounds.

What is their fur color?

Various shades of gray, with a distinctive brownish-black and white facial mask and tail.

What do they like to eat?

They are omnivores with an opportunistic diet and eat a wide variety of plant and animal foods. Favorite foods are bird eggs, fish and shellfish, frogs, birds, fruit, insects, nuts, and seeds.

What is their average life span in the wild?

2 to 3 years.

Fun Facts

- Raccoons are noted for their intelligence, and researchers have found that they are able to remember solutions to tasks for up to 3 years. Raccoons are considered smarter than cats and just below monkeys on the mammal intelligence scale.
- They make their dens in hollow trees, as well as abandoned burrows.
- They communicate with each other using over 50 different sounds and up to 15 different calls. They can purr, whistle, growl, scream, and hiss.
- They are known for rinsing their food in water prior to eating it. When no water is available, raccoons will still rub the food to remove debris. Their name Procyon lotor means, "washer dog."
- Raccoons are not related to dogs but are close relatives to bears.
- The name raccoon is derived from the Algonquin Indian word "arakun," which means, "he who scratches with hands."
- Raccoons are nocturnal animals, and one theory about their black mask around their eyes is that it helps the animals with their night vision.
- They have 5 fingers on their dexterous front paws, much like human hands, and are very skillful in opening shells, trash cans, doors, and collecting food. Their hands are so nimble they can unlace a shoe, unlatch cages, and even retrieve coins and small objects from pockets.
- They have the ability to climb down trees head first by rotating their hind feet 180 degrees.
- Raccoons are savage fighters and, if cornered, can kill an attacking dog.

Red Fox

Vulpes vulpes

What is their preferred habitat?

Forests and grasslands.

What do they like to eat?

They are omnivores but primarily feed on small rodents, birds, insects, and other smaller mammals. They readily eat plant material that includes berries, apples, grapes, grasses, and nuts.

How big are they?

Usually between 6 and 24 pounds.

What is their average life span in the wild?

2 to 4 years.

What is their fur color?

Orange-red fur on its head, back, and sides. White under its neck and chest, with a long bushy tail pointed in white. Black ears, legs, and feet.

Fun Facts

- Red foxes are great jumpers and can leap over 6-foot high fences.
- They are denning animals and dig burrows in which they live and raise families.
- Their bushy tails aid them in balance.
- They have vertical pupils and can retract their claws, two physical characteristics also of cats.
- They stalk their prey like cats, getting as close as they can and then pouncing. They have excellent hearing, especially for the low-frequency sounds produced by rodents digging underground.
- Female foxes are called vixens.
- Foxes have whiskers on their face and legs, which help them navigate.

Snowshoe Hare

Lepus americanus

What is their preferred habitat?
Coniferous forests with a dense brushy undergrowth.

How big are they?
They are larger than cottontail rabbits and typically weight between 3 to 4 pounds.

What is their average life span in the wild?
1 year or less.

What do they like to eat?
Grass, ferns, leaves, flowers, twigs, and bark. They will occasionally eat carrion.

What is their fur color?
White during winter months, and rusty to grayish brown during the rest of the year.

Fun Facts

- Snowshoe hares are rare in the Southern Appalachian Mountains, and what small populations that do exist are restricted to only a few high-elevation locations.
- They are solitary nocturnal animals and feed only from dusk to dawn.
- Their seasonal adaptation of fur turning white in the winter helps them to evade predators by blending in with the snow. Even though most of their fur turns white, the tips of the ears always remain black.
- Snowshoe hares are named because of the large size of their hind feet. These prevent them from sinking into the snow when they walk and hop.
- Snowshoe hares can jump almost 10 feet horizontally if necessary to escape predators.
- They also can run at speeds up to 25 miles per hour in short bursts. They are good swimmers and will also jump into ponds and streams to escape predation.

LEE JAMES PANTAS

Southern Bog Lemming

Synaptomys cooperi

What is their preferred habitat?

Grassy openings, bogs, swamps, and marshy areas in forests where the soil is moist and loose and covered with a thick mat of shrubby vegetation.

How big are they?

Small, usually weighing less than an ounce and about 5 inches in length.

What do they like to eat?

Leaves, stems, seed heads, roots of sedges and grasses, and berries.

What is their average life span in the wild?

1 to 2 years.

What is their fur color?

Reddish to dark brown with silver gray underbelly.

Fun Facts

- Southern bog lemmings are small voles that dig tunnels and runways for nesting, feeding, resting, and food storage.
- They are different from other voles in that they like to make small piles of grass next to their runways.
- Contrary to popular belief, lemmings are not suicidal and do not throw themselves off of cliffs.
- Lemmings are legendary for their population fluctuations and live in colonies that can number from a few up to 30.
- They excrete bright green fecal pellets due to the large amount of green vegetation they consume.
- Southern bog lemmings are most active at night, but also move about during the day.

Southern Flying Squirrel

Glaucomys volans

What is their preferred habitat?

Deciduous forests that have abundant nut trees.

How big are they?

Smaller than the Northern flying squirrel, around 8 to 10 inches.

What is their average life span in the wild?

4 to 5 years.

What do they like to eat?

They are omnivores and prefer nuts and acorns. They will also eat seeds, fungi, fruit, insects, eggs, birds, and carrion.

What is their fur color?

Gray brown on top and a cream color underneath.

Fun Facts

- Southern flying squirrels are nocturnal animals and are rarely seen by humans.
- They do not fly but instead glide using a membrane called a patagium that extends to either side of their body that helps to capture air. They have been observed gliding more than 150 feet.
- It is believed that they use triangulation to estimate the distance of the landing area when they launch themselves into the air.
- They nest in natural tree cavities and woodpecker holes, or build nests out of twigs and leaves.
- They are social animals and have been observed flying and foraging together in large groups.
- Size is one way to tell the Southern and Northern flying squirrels apart. The Southern flying squirrels average 8 to 10 inches, while the northern variety are usually 10 to 12 inches long.
- Humans imitate flying squirrels using special suits called "wingsuits," designed after the squirrel's anatomy with their extended flaps of skin. These wingsuits allow the experienced wearer, jumping from high mountainous elevations and off cliffs, to glide just like the squirrels over long distances.

Star-Nosed Mole

Condylura cristata

What is their preferred habitat?

Wet, poorly drained areas, marshes, and the banks of streams, lakes, and ponds.

How big are they?

Size ranges between 5 to 8 inches long and about 2 ounces.

What do they like to eat?

Invertebrates, aquatic insects, worms, amphibians, mollusks, and fish.

What is their average life span in the wild?

Not known for sure but probably less than 2 years.

What is their fur color?

Blackish brown.

Fun Facts

- The star-nosed mole is easily identified by the 22 pink fleshy appendages or rays that ring its snout and serve as a touch organ. There are more than 25,000 minute sensory receptors, known as Elmer's organs, in these rays, which help the mole feel it way as it digs underground.
- Most people would agree they are one of the world's strangest looking animals.
- They dig shallow tunnels for foraging that often end up underwater.
- They are excellent swimmers.
- They have been recognized by The Guinness Book of World Records as the world's fastest forager. This ability is due to their amazingly sensitive star nose that instantly tells the mole whether or not it has found food. It takes them as little as 230 milliseconds to identify and eat prey.
- They have a unique ability to smell underwater. They do this by exhaling air bubbles onto objects they encounter and then inhaling the scented bubbles back again.

Striped Skunk

Mephitis mephitis

What is their preferred habitat?

Woodlands, and brushy fields and meadows.

How big are they?

Usually between 2 to 14 pounds.

What is their average life span in the wild?

2 to 3 years.

What do they like to eat?

Primarily an insectivore, they will also eat vegetation, mice, voles, and the eggs and chicks of ground nesting birds.

What is their fur color?

Black with a white stripe extending back from the head, which divides the shoulders. Typically, they also have a white stripe on their foreheads.

Fun Facts

- The striped skunk is one of the world's most recognizable animals and is often featured in children's books and cartoons. Their black and white coloration is a warning to potential predators not to get too close.
- Like all skunks, they possess scent glands under their tail. The glands contain an oily, yellow, foul smelling musk that is sprayed when threatened by predators.
- Skunks are easily tamed, kept as pets, and also have been extensively bred for their fur.
- Skunks sometimes dig their own dens, but will readily use those abandoned by other animals. They also live in old tree stumps, under logs and rocks, and even in attics and basements of houses.
- In the 19th century, skunks were often kept in barns by farmers to kill mice and rats.
- They are nocturnal animals and hunt for food only at night.
- Skunks are very intelligent animals, and one of their more interesting behaviors is their practice of rolling a furry caterpillar in the dirt to get rid of its irritating hairs before eating it. They also can easily get into garbage cans, will eat pet food given the opportunity, and are known to raid vegetable gardens and compost bins.
- They are more likely to be hit and killed by a car than killed by natural predators.

Virginia Opposum (Possum)

Didelphis virginiana

What is their preferred habitat?

Wet areas along streams, and marshes and swamps.

How big are they?

About the size of a house cat, usually between 10 and 14 pounds.

What is their average life span in the wild?

1-2 years.

What do they like to eat?

They are omnivores, and eat a wide variety of foods including fruit, grain, insects, earthworms, and small mammals.

What is their fur color?

Gray on their bodies with a whitish face and dark ears.

Fun Facts

- The Virginia opossum is the only marsupial found in North America, and has been in existence for at least 70 million years.
- They have been known to have as many as 25 babies, but the typical litter is between 7–8. Their babies, once born, climb into the mother's pouch where they find teats for milk.
- During the autumn months, persimmons are one of their favorite foods.
- Possums are known for reacting to threats by pretending to be dead, thus discouraging attacks by predators. This reaction is involuntary and triggered by extreme fear. If overly threatened, they will enter a near coma and lie on their side, mouth and eyes open, tongue hanging out, and their breathing rate slowing by 30% or more. This death feigning normally stops when the threat is withdrawn but can sometime last up to 4 hours.
- They have a prehensile tail that is adapted for grasping and wrapping around tree limbs.

White Squirrel

Sciurus carolinensis

What is their preferred habitat?

Forests and woodlands, preferably those with oak, walnut, hickory, and other nut-bearing trees.

What do they like to eat?

Acorns, nuts, seeds, tree buds and bark, and food offerings in residential bird feeders.

How big are they?

1 to 2 pounds.

What is their average life span in the wild?

8 to 9 years.

What is their fur color?

White

Fun Facts

- White squirrels are color variations of the gray squirrel, not a separate species or albinos.
- Evidence that they are simply a color variation is seen in their protective behavior, demonstrated when they are foraging on the ground and then threatened by a predator or human. When that happens, they dart to the nearest tree trunk and immediately flatten themselves against the tree trunk to avoid being seen. This is appropriate for a squirrel with a gray color as a way of camouflaging itself (gray fur on gray tree), but a white squirrel stands out highly visible against the darker tree bark.
- They are found in the Southern Appalachian Mountains in and around Brevard, North Carolina, just west of Asheville in the western part of the state.
- Brevard has a yearly "White Squirrel Festival" each May, named after these animals.
- There are four other cities in America and Canada that also have resident populations of white squirrels. These are Olney, Illinois; Kenton, Tennessee; Marionville, Missouri; and Exeter, Ontario, Canada.
- White and gray squirrels are scatter-hoarders, animals that store food in numerous caches for later recovery during winter months. Individual squirrels can make up to several thousand caches each season.
- Squirrels sometimes use deceptive behavior to prevent other animals from finding their food caches by pretending to bury a food item when they think they are being watched.
- White and gray squirrels are one of the few mammals that can descend a tree head first.
- They construct their dens as nests from twigs and leaves high up in tree branches, or within the hollow trunks of trees.

White-Tailed Deer

Odocoileus virginianus

What is their preferred habitat?

During summer, fields and meadows with bordering oak and hickory forests for shade. During the winter, preferably forests with coniferous stands that provide protection from the elements.

How big are they?

Usually between 100 to 150 pounds, but can weigh up to 300 pounds and grow to over 7 feet in length.

What do they like to eat?

They are herbivores, with a diet of grass, leaves, twigs, fruit, and nuts.

What is their average life span in the wild?

3 to 5 years.

What is their fur color?

Reddish brown in the summer and duller gray-brown in the winter.

Fun Facts

- Male deer are called "bucks," females are known as "does," and young deer as "fawns."
- Fawns are born with white spots as camouflage, but lose these within a year of birth. Fawns also take their first steps with 30 minutes of their birth. Typically, fawns stay with their mothers for about one year.
- They can run up to 45 miles per hour in bursts and leap as high as 9 feet, easily clearing any fence.
- They have four-chambered stomachs that allow them to digest tough plant foods.
- Their white tails are highly visible when they are running.
- They are also excellent swimmers and often enter water to escape predators, or to reach new feeding grounds.
- White-tailed deer bucks grow horns that are grown annually and fall off in the winter. With each passing year, the newly grown antlers will have more points. The bucks use their horns as defensive weapons and in mating battles with other bucks.
- When alarmed, they stomp their hooves and snort to warn other deer. They may also "flag" or raise their white tail as a signal.
- Deer are part of the Cervidae family, which include reindeer, elk, and moose.

Wild Boar

Sus scrofa

What is their preferred habitat?

Dense vegetation and woodlands near accessible water.

How big are they?

Usually between 100 to 200 pounds but can grow to over 400 pounds.

What is their average life span in the wild?

1 to 2 years.

What do they like to eat?

They are omnivores and prefer crops, fruit, nuts, roots, and green plants. They also will consume bird eggs, small wild mammals, young livestock, and carrion.

What is their fur color?

Brown, dark gray or reddish brown, with many variations.

Fun Facts

- The wild boar is a species of wild pig native to the forests of Europe, Asia, and parts of Africa that was introduced into America.
- They have extremely poor eyesight, compensated for by an incredibly acute sense of smell and excellent hearing.
- They live in large groups, called sounders, made up of 6 to 20 closely related females and young. These groups can have as many as 100.
- They are considered an invasive species in the Southern Appalachian Mountains, as well as other areas in North America. Their habit of rooting and digging destructively in the soil, and raiding farm crops for food, are the main complaints about wild boars.
- Male wild boars can have razor sharp tusks as long as 5 inches that they use for defense, or in establishing dominance over other males.

Woodland Jumping Mouse

Napaeozapus insignis

What is their preferred habitat?

Woodlands and forests with cool moist areas and dense vegetation.

How big are they?

Smaller than their cousins the meadow jumping mouse, and usually between 8 to 10 inches long from the tip of the nose to the end of their long tail.

What do they like to eat?

Seeds, fruit, fungi, nuts, leaves, and insects.

What is their average life span in the wild?

3 years.

What is their fur color?

White on the belly, dark yellowish-orange to reddish-brown on the sides, and a darker brown stripe down their back.

Fun Facts

- Woodland jumping mice shelter in burrows they dig, and are normally active during the night, but also can be seen on cloudy or rainy days. When they are in their burrows, they typically cover the entrances with leaves or other vegetation. Their burrows are lined with grass, reeds, and leaves.
- They can swim both underwater and on the surface of water for short distances.
- Their lifespan, up to four years, is among the longest for small rodents and mammals. This is partly due to their ability to hibernate. They hibernate at least 6 months of the year, from fall until spring.
- Normally they walk using all four legs but when speed is needed, they resort to a hopping motion. When they are frightened, these hops can be almost 9 feet long horizontally and as high as 2 feet in the air. Long hind legs with elongated ankle bones, long toe bones, and long tails, make it possible for them to be such extraordinary leapers. One observer has compared them to miniature kangaroos when they jump.
- Their long tails are designed to help them with balance when jumping and hopping.

Field Notes and Observations

Field Notes and Observations

Field Notes and Observations

Made in USA - North Chelmsford, MA
1123923_9780991039821
09.04.2020 1617